Physics Lab Guide

Motion, Force, Newton's Laws, Torque, Energy, Heat, and Buoyancy

Elmar Bergeler, Ph.D.

Texas State University – San Marcos

© 2019 Science and Math Publishing

All rights reserved. No part of this publication may be reproduced, distributed, or transmitted in any form or by any means, including photocopying, recording, or other electronic or mechanical methods, without the prior written permission of the publisher.

ISBN-13: 9781097441990

Topics:

Motion, force, Newton's laws, torque, energy, heat, buoyancy

Manuals about:

Logger Pro software, LabQuest 2 interfaces, motion detectors, temperature sensors, photogates, force sensors (Vernier)

ImageJ/Fiji, Excel

Content

Introduction 7

Part A

Measurement Principles, Methods, and How to Present Data

1. Taking Measurements 11
    1.1. Controlling Variables 11
    1.2. Collecting Data 11
    1.3. Units 12
            Prefixes 14
    1.4. Measurement Tables 15
    1.5. Measurement Graphs 16

2. Accuracy, Uncertainty, and Measurement Error 19
    2.1. Uncertainty and Measurement Error 19
            Standard Deviation and Measurement Error 19
            Consistent Measurements 20
            Percent Error 20
            Error Propagation in Calculations 20
            Error Types and Sources 21
            Reading Displays and Errors 22
    2.2. Significant Figures 22
            Using the Correct Number of Significant Figures 23
            Significant Figures in Calculations 25
            Significant Figures of Errors 26

Part B
Manuals and Instructions for Laboratory Equipment and Software

3. Using LabQuest 2 and Logger Pro 29
    3.1. Getting Ready to Take Measurements 29
    3.2. Setting up Sensors Manually 30
    3.3. Motion Detector 31
        Hardware 31
        Software 31
    3.4. Force Sensor 31
        Hardware 31
        Software 32
    3.5. Photogates 32
        Hardware 32
        Software 33
        Using a Premade Program to Work with Photogates 33
    3.6. Temperature Sensor 35
        Hardware 35
        Software 35

4. Using ImageJ/Fiji for Video Analysis 37
    4.1. Install ImageJ/Fiji 37
        Add the Plug-in for iPhone and Android phone videos 37
    4.2. Start ImageJ/Fiji 38
    4.3. The Motion Videos 38
        Recording our own Videos 38
    4.4. Import a Video into ImageJ/Fiji 39
        Rotating the Video 41
    4.5. Set up a Scale 41
    4.6. Data Collection 42

Show Motion Path 44

Create Motion Data with ImageJ 44

Analysis of Motion Data from the Video 45

5. Creating Measurement Graphs with Excel 49

5.1. Creating new Variables and Quantities Using Formulas 49

5.2. Creating Graphs with Excel 50

Adding Labels and Modifying the Graph 52

5.3. Adding Regression Lines and Curves in Excel 52

Linear Regression 53

Quadratic and Exponential Regression Curves 53

6. Creating Measurement Graphs with Logger Pro 57

6.1. Zoom in or out in the Graphs 57

6.2. Adding Regression Lines and Curves in Logger Pro 57

Linear Regression 57

Quadratic and Exponential Regression 58

7. Data Analysis 59

7.1. Clean up the Measurement Data 59

7.2. Averages 59

7.3. Comparing the Result with the True Value 60

7.4. Measurement Errors 60

7.5. Analyzing Graphs 61

Linear Regression 62

Quadratic Regression 62

Exponential Regression 64

Accuracy of the Regression Lines and Curves 64

Part C

Lab Experiments

> Lab 1: Measure and Describe Motions 69
> Lab 2: Velocity and Acceleration 71
> Lab 3: Two- and One-Dimensional Motion Using Video Analysis 75
> Lab 4: Forces 79
> Lab 5: Newton's Second Law of Motion 83
> Lab 6: Static and Kinetic Friction 87
> Lab 7: Rotational Equilibrium and Torque 91
> Lab 8: Momentum and Collisions 95
> Lab 9: Kinetic and Potential Energy Transformations 99
> Lab 10: Specific Heat 101
> Lab 11: Buoyancy 105

Formulas and Constants **108**

Introduction

This lab guide provides students with the basic knowledge needed to successfully participate in an algebra-based physics laboratory course. This guide is an ideal addition to any introductory physics text. This book guides students through hands-on experience with computer-based experiment equipment, video analysis of motions, and real-world applications of physics concepts. This lab guide gives step-by-step instructions about how to use a Vernier measurement system consisting of the software Logger Pro, the LabQuest 2 interface, and the most common sensors. This guide also explains how to use the video analysis program ImageJ/Fiji to take measurements. However, the experiments in this guide leave room for their own thoughts, activities, and experimental designs, so that students learn experimental skills. Through this guide, students also learn how to create measurement graphs with Microsoft Excel, how to analyze measurement data, and how to analyze measurement errors.

This lab guide has three parts. Part A focuses on measurement theory, methods, how to present data, and units. Part B contains manuals for common experimental laboratory equipment and software used in physics lab courses. Part C is a collection of lab experiments about motion, force, Newton's laws, torque, energy, heat, and buoyancy. The lab experiments are designed for a weekly two-hour lab course which goes along with the algebra-based lecture for General Physics 1.

The physics theory and concepts covered by the experiments in this book are about are explained in the textbooks which are used in the lecture that goes along with the lab course. Another resource for the physics content is the textbook

College Physics by Paul P. Urone, Roger Hinrichs, Kim Dirks, and Manjula Sharma.

It can be downloaded for free from the OpenStax CNX website at

https://openstax.org/details/books/college-physics

Contents of Part A

Chapter 1 is about measurements and units.

Chapter 2 is about measurement errors and significant figures.

Contents of Part B

Chapter 3 gives step-by-step instructions about how to use a Vernier measurement system consisting of the software Logger Pro, the LabQuest 2 interface, and the most common sensors.

Chapter 4 is about the video analysis program ImageJ/Fiji.

Chapter 5 is about how to create graphs from your data using Excel.

Chapter 6 is about how to create graphs from your data using Logger Pro.

Chapter 7 is about data analysis.

Contents of Part C

Part C is a collection of experiments using Vernier's measurement systems and the video analysis software ImageJ/Fiji with pre lab assignments. The topics of the experiments are motion, force, Newton's laws, torque, energy, heat, and buoyancy.

Part A

Measurement Principles, Methods, and How to Present Data

1. Taking Measurements

1.1. Controlling Variables

When we take physical measurements, we should work as accurately as possible or required. We must control the environment of the experiment in order to eliminate factors that disturb our measurements. To find out how one or more variables depend on another variable, we do physics experiments in which we change one quantity systematically from small to large values or vice versa and observe how other variables depend on the one we change. While we change the one variable systematically, we keep all other quantities constant. To be able to replicate our results, we record all data.

1.2. Collecting Data

When we take a measurement of a physical quantity, it always consists of a number and a unit (see section 1.3).

To investigate a possible relationship between different physical quantities we take a whole series of measurements. We change one quantity, which we call the independent variable, in a series of measurements and observe the value of the respective dependent quantity. We record both values in a measurement table (see section 1.4). The measurement data comes in pairs of the two variables, which are the measurement points. Often, we present these measurement points in graphs (see section 1.5). We want to collect at least 5 to 10 data points to find out if and how these quantities are correlating.

The measurements can be taken manually, where we take each measurement with our device one by one. However, very common are automatically gathered measurement data using a computer-based measurement system (see chapter 3). Or we can use video analysis software (see chapter 4) to take position data, which can also be done one by one, or we can have the video analysis software track an object automatically.

Depending on what type of device or hardware we use to take the measurements, each measurement comes with a certain measurement error and accuracy (see chapter 2). When we record the measurement values, we must not record the values with more accuracy than the device or method allows.

1.3. Units

Measurement values consist of a number and a unit. For example, we can measure a distance to be 123 feet. In this case "feet" is the unit we used to express the distance. Possible other units of length are miles, inches, meters, kilometers, and many more. Mostly two sets of units are used: the Imperial System (also called British Imperial System or English System), and the International System of Units (called SI for short). Its original name is in French: *Système Internationale d'Unités*. Its units are called SI units.

The SI is a metric system. That means when we switch from smaller units to larger units of the same quantity or vice versa, we only move the decimal point to the left or to the right (which means dividing or multiplying the value by a power of ten). Prefixes before the units are used to change the magnitude of the units.

The Imperial system is mostly used in the USA, Great Britain, and some former British colonies in everyday life. The SI units are used in everyday life everywhere else. However, in science the SI units are used throughout the world, including in the USA.

Therefore, we want to use the SI units in physics classes. If we use a physics formula to calculate any quantity, we must use SI units when we plug in the values. Then, the result is automatically in the correct SI unit for that quantity. If we start with non-SI units, we first must convert the values to SI units before we start the calculation.

The table below shows all basic SI units. The basic SI units don't have any prefixes, except for kilograms.

Quantity	Basic SI Unit	Abbreviation
Time	Second	s
Length	Meter	m
Mass	Kilogram	kg
Electric current	Ampere	A
Temperature	Kelvin	K
Amount of substance	Mole	mol
Luminous intensity	Candela	cd

All basic SI units

The units for any other quantity can be derived from the basic SI units. The derived units are still SI units, just not basic SI units. Some derived SI units which are commonly used in physics labs are shown in the table below.

Quantity	SI Unit	Abbreviation
Speed or Velocity	Meters per second	m/s
Acceleration	Meter per second squared	m/s²
Force	Newton	N or $\frac{kg\,m}{s^2}$
Energy	Joules	J or $\frac{kg\,m^2}{s^2}$
Power	Joules per second or Watts	J/s or W
Torque	Newtonmeter	Nm
Pressure	Newton per square meter or Pascal	N/m² Pa

SI units for some more quantities.

Prefixes

Prefixes are used to change the magnitude of units by a factor of a power of ten. Prefixes allow writing very large or very small values more compactly.

The units with prefixes are still considered SI units (as long as they are based on an SI unit). The table below shows some common prefixes. The prefixes can be used in front of any unit to change its magnitude.

Prefix	Symbol	Factor	Example
nano	n	10^{-9}	12 nm = 12×10^{-9} m = 0.000 000 012 m
micro	μ	10^{-6}	34 μm = 34×10^{-6} m = 0.000 034 m
milli	m	10^{-3}	25 mm = 25×10^{-3} m = 0.025 m
centi (only for length)	c	10^{-2}	7 cm = 7×10^{-2} m = 0.07 m
deci (only for length)	d	10^{-1}	3 dm = 3×10^{-1} m = 0.3 m
kilo	k	10^{3}	44 km = 44×10^{3} m = 44 000 m
mega	M	10^{6}	Not common for length, but bytes (b): 5 Mb = 5 ×10^{6} b = 5 000 000 b
giga	G	10^{9}	Not common for length, but bytes (b): 37 Gb = 37 ×10^{9} b = 37 000 000 000 b

Table with common prefixes

> **Example**
>
> The basic SI unit for length is meter. But when we deal with small lengths, it is not convenient to use just the unit meter. Instead we can use micrometer, millimeter, or other units for lengths. For these units, meter has a prefix to change the magnitude of the unit. E.g. instead of using meters we use micrometers if the length is very small:
>
> 0.000 05 m = 50 μm

> **Attention!**
>
> When we use a physics formula and plug in values and want the result to be in an SI unit, all values we plug in must not have any prefixes, except for kilogram (kg).

1.4. Measurement Tables

Typically, measurement values from experiments are recorded in a table (see example below). This gives us a good overview of the relationship between the quantity we change (called the independent variable) and another quantity which depends on it (called the dependent variable). When we create the table manually, we must not forget to label the first column with the variable names we record. Also, we must not forget to give the units, preferably SI units. When we use a computer-based measurement system, the table is created automatically.

Tables are also necessary to create measurement graphs. Tables can be oriented horizontally, like the one in the example below, or vertically.

> **Example**
>
> To find out how the acceleration depends on the applied force, we change the accelerating force from 0 Newton to 100 Newton step-by-step and record the respective acceleration for each force value. We want to keep all other variables, except the force, constant. That means we don't change the mass or anything else from your experimental setup at the same time. Our measurements should be reported in a table like the one below. The first column shows the quantity and the unit. The slash "/" means "in" and indicates what units we are using for our measurements.
>
> | F / N | 0 | 0.196 | 0.392 | 0.588 | 0.784 | 0.980 |
> |---|---|---|---|---|---|---|
> | a / m/s² | 0 | 0.43 | 0.78 | 1.22 | 1.46 | 1.99 |

1.5. Measurement Graphs

Another typical way of representing data is as a graph. Before we can create the graph, we need the measurement table. Usually, the independent variable (the one we change through the experiment or the time) is plotted on the horizontal axis. The dependent variable (the one which depends on the independent variable) is plotted on the vertical axis.

For example, if we alter the force and observe the resulting acceleration, we would plot acceleration on the vertical axis and the force on the horizontal axis. We can also say the acceleration is plotted over the force. Such graph is also called an acceleration-versus-force graph. The first word tells us what's on the vertical axis, and the second tells us what's on the horizontal axis.

When we create a graph, we pay attention to the following:

- The distances between the tic marks on each axis must be equal.
- The axes start at zero, and their values increase upwards and to the right.
- Label the axes with what is plotted.
- Add units to the axes' labels.
- If the values can change continuously, even when you only have a couple of distinct measurements, add a regression line or curve, which best fits your data.
- Use the whole graph area to display the data of interest (the graph must be zoomed in or out respectively).

Tip:

When you create a graph with Excel, or directly in the measurement software, you have two easy ways to export it for your lab report:

- You can use the copy-and-paste function after clicking on the graph.
- You can take a screenshot of the graph and crop it in Paint and then save it as a graphic file, which you can insert in your report.

Example

The measurement graph for the data:

F / N	0	0.196	0.392	0.588	0.784	0.980
a / m/s²	0	0.43	0.78	1.22	1.46	1.99

2. Accuracy, Uncertainty, and Measurement Error

No measurement is 100% accurate. Every measurement comes with measurement errors. Even the most accurate measurements come with an uncertainty because of that error.

2.1. Uncertainty and Measurement Error

Because of the measurement error, each value we measure or calculate comes with an uncertainty. The largest possible error determines the uncertainty interval. The actual error could be smaller. But the problem is that we don't know it exactly. When we measure a value, then the true value is somewhere within the error interval, as shown in the figure below. The true value cannot be measured.

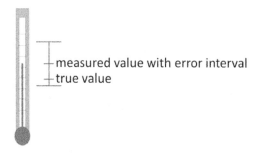

Figure of measurement value with error interval

Standard Deviation and Measurement Error

The measured value can be larger or smaller than the true value. The true value of the quantity x_{true} cannot be directly measured; however, we know it must be close to the value we measure. The true quantity x_{true} is then given by

$$x_{true} = x_{measured} \pm SD$$

where $x_{measured}$ is the measured value of that quantity and SD is the standard deviation.

The standard deviation spans the uncertainty interval around our measured value. It indicates the measurement error, how much the values scatter around the true value.

Every device gives the measurement value with an uncertainty interval. This depends on the electronics or mechanics of the device. High-end lab equipment is labeled with the measurement error of the device.

Consistent Measurements

Since every measurement comes with an uncertainty, we often have different measurement values, even though they represent the same value. Thus, we look at the whole measurement value as well as the uncertainty interval when we compare measurement values. If the intervals of different measurements

$$x_{measured} \pm SD$$

are within each other's range, we say the measurements are consistent with each other. SD is the standard deviation for the measurement $x_{measured}$.

> **Example**
>
> When we have two temperature measurements
>
> $$T_1 = 281K \pm 2K$$
>
> and
>
> $$T_1 = 284K \pm 1K$$
>
> We see that these values are within range, because their uncertainty intervals overlap.
>
> We say these values are consistent with each other.

Percent Error

The error can also be given as percent error (%$error$), by what percentage the measured value is possibly larger or smaller than the true value:

$$\%error = \frac{SD}{x_{measured}} \times 100\%$$

The true value for the quantity x_{true} is then given by

$$x_{true} = x_{measured} \pm x_{measured} \times \frac{\%error}{100}$$

Error Propagation in Calculations

When we derive further values from measurements with measurement errors, the error propagates through the calculations and is passed on to the result:

- When we multiply or divide two numbers which each have a percent error, then the result's percent error is

$$\%error = \%error_1 + \%error_2$$

- When we add or subtract two numbers which each have a standard deviation, then the result's standard deviation is

$$SD = SD_1 + SD_2$$

> **Example**
>
> When one of our measurement values has a percent error of 5%, and another one has an error of 3%, then, if you multiply these two measurement values, the result has a percent error of 8%.

Error Types and Sources

Measurement errors originate from the tools and devices, calibration, or the method, by which the measurement is performed. Or errors can also just originate from statistical scattering of the noise of the devices, or unknown influences from the surroundings, which can't be controlled.

There are two types of errors: random errors and systematic errors.

Random Error

A random error means that the measurement is randomly off within a certain range around the true value. This creates the uncertainty interval as discussed in section 2.1. Our measurement values can be unpredictably higher or lower than the true value. When we repeat measurements, we always observe a scattering around the true value.

Systematic Error

A systematic error is, if the measurement is always offset by the same value. For example, if we forget to calibrate the force sensor to 0 N when no force is applied; maybe it already shows 0.043 N when no force is applied. Then all our measured values are 0.043 N too high.

Reading Displays and Errors

When the measurement value is shown on a digital display, by default at least the last digit is uncertain. The actual value is always within ±1 of the last displayed digit. This is because the rounding method of that device is unknown. We don't know if it rounds correctly or if it just cuts off the value at that decimal place. Analog displays have an uncertainty of at least of ± ½ of the last readable decimal place of the measurement value by default.

2.2. Significant Figures

Significant figures in a number are figures of significance. That means the decimal places of a number we keep and don't round reflect the accuracy of that value. The table below shows how to determine the number of significant figures of a given number.

Type of number	Rule	Example
Integers like 3007 or −3007	When an integer starts and ends with a figure other than 0, then every figure in that number is significant.	3007 and −3007 each have 4 significant figures.
Integers like 7500 or −7500	When an integer ends with 0s, then all zeros at the end are not significant.	7500 and −7500 each have 2 significant figures.
Decimal numbers like 120.40 or −120.40	If a decimal number is greater than 1 or less than −1, then all figures are significant. You can see it this way: The last 0 of that decimal number is written for the only reason, to express a higher accuracy of that number than without that 0.	120.40 and −120.40 each have 5 significant figures.
Decimal numbers like 0.000450 or −0.000450	If the absolute value of decimal number is smaller than 1, then all zeros before the non-zero places are not significant. However, the trailing zeros are significant.	0.000450 and −0.000450 each have 3 significant figures. In contrast, 0.0004500 has 4 significant figures and 0.00045 has 2 significant figures.

Table showing how to determine the number of significant figures

Using the Correct Number of Significant Figures

How many significant figures for our measurements should we record or keep in the result, before rounding?

How many decimals we use before we round depends on the uncertainty of the measurement value or the values we used for our calculations.

The number of figures we use is another simplified way to express accuracy, instead of using the actual uncertainty interval (see section 2.1 above). The uncertainty is ± ½ of the last significant decimal place of a value if no uncertainty interval is given. The number of significant figures should match the accuracy. Don't give results with more decimal figures than the uncertainty allows: The smallest decimal places in a result must not be smaller than the error.

Not all figures we see on a display, or that come out of a calculation, are actually significant. If a value for one measurement fluctuates continuously, then the changing figures are not reliable and thus not significant. We never give more significant figures than the margin of error allows. We round values from a measurement and give less accuracy than on the display, or we round results from a calculation in the following situations:

- Our measurements aren't that accurate anyway. If we give too many figures, then the number implies a higher accuracy than it actually has.
- The accuracy doesn't matter, and rounded values look nicer, or are easier to deal with.
- The result of our calculation contains more figures than our original number we plugged into the formula, because of the math. But when we report our final answer, we must use the same accuracy (or less) as the original values we used for our calculation. (See the section below about significant figures of results from calculations.) We reduce the accuracy in the result by rounding. The only exception when the accuracy improves is when we calculate average values (see also section 7.2 for carrying out data analysis).

> **Example**
>
> If a value on the display for a force measurement shows the measurement value to four decimal places but the last two digits of the value fluctuate, we report the measurement value with less accuracy than the display shows. For example, if the reading of a force is 2.7842 N, but the last two digits fluctuate during one single measurement, then we round the measurement value to 2.78 N.

Significant Figures in Calculations

When we do calculations with values which have different numbers of significant figures, we need to determine how many significant figures our result has. The rules are:

- When multiplying or dividing two numbers, the number of significant figures of the result is equal to the number of significant figures of the number with the fewest significant figures.
- When adding or subtracting two numbers, the number of significant figures of the result is determined by the number with the least accuracy. The result's smallest decimal place matches the smallest decimal place of the two added or subtracted numbers with the larger smallest decimal.

> **Example**
>
> 12.4 m × 1.2 m = 14.88 m^2
>
> But you should give the result as 15 m^2, because 1.2 m has only two significant figures.
>
> 321 m + 4.3 m = 325.3 m
>
> But you should give the result as 325 m, because the smallest decimal place of 321 m is full meters, so the result can also only be given to the accuracy of full meters.

> **Attention!**
>
> In interim calculations we should keep more decimal places than in our final result to prevent rounding errors.

Significant Figures of Errors

When we report the uncertainty in our final result, we should only use one significant figure for the error. We must round up the measurement value to the same decimal place as the uncertainty or error.

Example
m = 13.453 kg ± 0.012 kg becomes
m = 13.45 kg ± 0.02 kg

Attention!
The error must always be rounded up.

Part B

Manuals and Instructions for Laboratory Equipment and Software

3. Using LabQuest 2 and Logger Pro

The measurement procedures in this lab guide are explained for LabQuest 2, which is connected to a computer with the Logger Pro software version 3.14.1. Both products are produced by Vernier. If you use another version, the procedures and menu items might be slightly different.

3.1. Getting Ready to Take Measurements

In a lot of physics labs, the measurements are collected using the software "Logger Pro" in combination with the device "LabQuest 2." There is usually a Logger Pro icon on the desktop which we double-click to start the program (see figure below).

Logger Pro
3.14.1

The LabQuest 2 is connected to the computer via USB. The individual sensors are connected to the LabQuest 2, and most sensors are automatically detected from the device and are instantly ready to be used. The LabQuest 2 has two types of inputs for the sensors:

- Analog inputs "CH 1," "CH 2," and "CH 3"
- Digital inputs "DIG 1" and "DIG 2"

Depending on the sensor type, the sensor can only be connected either to one of the analog inputs or to one of the digital inputs.

If the LabQuest 2 is connected to the computer, we can sample the data directly with the Logger Pro software. We can also create graphs and analyze the data within the measurement software. But we can also sample the data with just the LabQuest 2, without having it connected to the computer, for mobile data collection. In this case the measurement data will be stored on the LabQuest 2 itself, and the next time we connect it via USB to the computer with the Logger Pro software running, it will prompt us to import the data from the LabQuest 2 to the computer.

While the device is taking measurements, the data is shown in a measurement table and in a graph in real time by default.

Section 3.2 describes how to set up sensors in Logger Pro manually, in case they are not recognized and set up automatically. However, normally we shouldn't have to do this. Sections 3.3 – 3.6 show how to use the most common sensors to take measurements for the physics lab.

3.2. Setting up Sensors Manually

If a sensor is not automatically detected and set up by the software, we must do that manually. To set up a sensor, we follow these steps:

- We click on the "Sensor Setup" symbol of the Logger Pro window, circled in the figure below. This opens up the sensor setup menu.

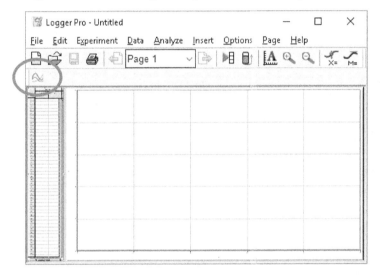

- In this menu we choose the sensors we want to use for the individual inputs (see figure below).

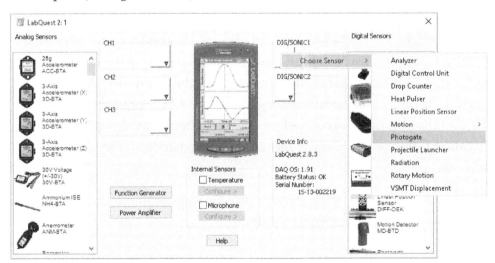

When we are done selecting sensors, Logger Pro automatically creates the table with columns for the measurements the selected sensors deliver.

3.3. Motion Detector

We use the motion detector from Vernier to measure the position and velocity of an object that moves in a straight line.

Hardware

We connect the motion detector to the LabQuest 2 on port "DIG 1" or "DIG 2." The motion detector detects the distance to an object via ultrasonic signals. There is a minimum distance of a few centimeters (or a few inches) and a maximum distance of around 8 meters (or 25 feet). Also, we make sure the sonic signal hits the right target to measure the correct distance and not some unintended distances to other objects nearby or to the walls.

The motion detector has a switch to choose between two settings, which depend on the experimental setup.

- Ball/Person setting:
  If we want to measure the distance to a small ball, person, or any other object of that size or bigger.
- Cart setting:
  If we want to measure the distance to a small cart on a track.

Software

The LabQuest 2 device automatically recognizes the motion detector and sets all necessary measurement settings as soon as the motion detector is connected. The device is immediately ready to take motion data.

To start taking measurements, we click on the green triangle in the upper right corner of the Logger Pro window.

3.4. Force Sensor

We use the force sensor to measure forces which pull or push on the device.

Hardware

We connect the force sensor to the LabQuest 2 on port "CH 1," "CH 2," or "CH 3." The sensor measures the force with which the hook is pulled or pushed. The force sensor has a sensitivity setting. Depending on the force we want to measure, we can choose between ± 10 Newton or ± 50 Newton to get the best sensitivity, using the switch on the force sensor.

Software

The LabQuest 2 automatically recognizes the force sensor and sets all necessary sensor and measurement settings as soon as the sensor is connected. The device is immediately ready to collect force measurements.

To start taking measurements, we click on the green triangle in the upper right corner of the Logger Pro window.

> **Attention!**
>
> Don't forget to calibrate the force sensor before beginning your measurements. Set the force sensor to zero when no force acts on it by clicking on the "∅" symbol in the upper right corner of the menu bar of the Logger Pro window.

3.5. Photogates

We use photogates to measure speeds of objects when they pass the photogate.

Hardware

The photogates only work if the LabQuest 2 is connected to the computer with the Logger Pro software. We can take measurements with one or two photogates simultaneously. The photogates work with infrared signals (IR- signals). The actual measurement consists of recording whether the gate is blocked or unblocked at a given time. All other measurement values, such as velocities, are calculated values which are derived from the detected states and their times.

We connect the photogates to the LabQuest 2 on port "DIG 1" or "DIG 2." The photogates are usually used together with carts which run through the gates and block the IR- signal. On top of the carts needs to be a strip which blocks the photogate's IR- signal when the cart runs through it. Other objects can be used in place of the carts, as long they fit through the photogate and can block the IR- signal.

Software

To take measurements with the photogates, the LabQuest 2 needs to be connected to the computer with the Logger Pro software running. LabQuest 2 and Logger Pro don't recognize the photogates automatically. Before we can take measurements, we must manually set up the photogate sensors and measurement quantities. The photogates only output the weather the gate is blocked or unblocked. From these states and their times, we can derive other quantities, such as velocity. To get any other output using the photogates, we need to create and define these quantities. To make Logger Pro output these quantities, we have to enter the respective parameters and formulas which the software uses to calculate these outputs.

Using a Premade Program to Work with Photogates

Because of the complications with the photogates, I recommend using one of the factory-supplied programs to save the hassle of manually setting up the sensors and measurement quantities such as the velocity.

The factory-supplied program "Two Gate Timing.cmbl" already contains all settings to take time and velocity measurements with two photogates. But we can also use this program if we want to take measurements with only one photogate.

To open "Two Gate Timing" in Logger Pro:

- Under the "File" menu choose "Open…" in the submenu.
- Navigate in the file browser to "Probes and Sensors" → "Photogates" → "Two Gate Timing."

After opening the file "Two Gate Timing" the screen should look like the figure below.

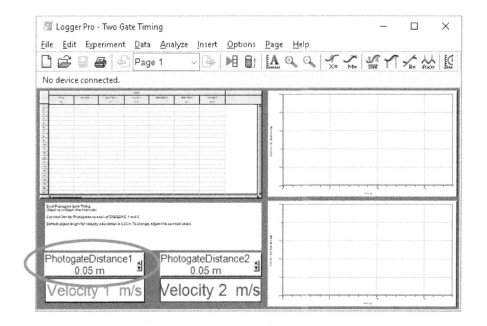

> **Attention!**
>
> If our strip which blocks the photogate is not 5 cm long, we still have to set the correct length of the strip in the software. Otherwise the measured speed is wrong.

The velocity is calculated through the time the photogate is blocked by the black strip, and the length of the strip. To set the length of the strip we:

- Double-click on "PhotogateDistance1" to change the size of the black strip (see circled field in the image above).
- In the next window (see figure below) enter the size of the strip which blocks the photogates in the "Value" field. Enter the number of decimals in the "Places" Field.

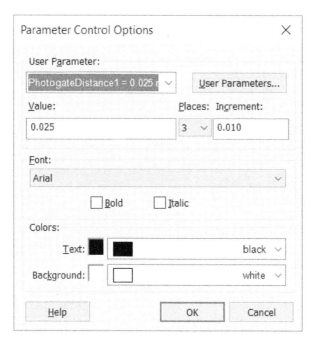

- After clicking "OK," we are ready to take velocity measurements at the photogates.

To start taking measurements with Logger Pro, we click on the green triangle in the upper right corner of the Logger Pro window.

3.6. Temperature Sensor

We use the temperature sensor to measure the temperature of an object or a liquid which touches the sensor's metal rod.

Hardware

We connect the temperature sensor to port "CH 1," "CH 2," or "CH 3" of the LabQuest 2.

Software

The LabQuest 2 automatically recognizes the temperature sensor and sets all necessary measurement settings as soon as the sensor is connected. The device is immediately ready to take temperature data.

To start taking measurements with Logger Pro or the LabQuest 2 alone, we click on the green triangle in the upper right corner of the window.

To change the temperature unit to Kelvin, we go to the top menu "Experiment" in Logger Pro and choose "Change Units."

4. Using ImageJ/Fiji for Video Analysis

ImageJ is a video analysis software with which we can collect motion data from a video. If we have a video which shows the motion of an object, we can take position-versus-time data pairs using the time and pixel information of that video. In this chapter I will show how to do manually track motions in a video. However, ImageJ even lets you track objects automatically.

I recommend the Fiji distribution from ImageJ, since Fiji supports all common video formats. I use the Windows version in this lab guide.

If ImageJ/Fiji is used in a lab course, it is probably already installed on the lab computer and has an icon on the Windows desktop.

4.1. Install ImageJ/Fiji

We can download ImageJ/Fiji for free at

https://fiji.sc/#download

The downloaded file from the above link is a zipped folder. We must extract it at a location where we have full write access (e.g. the Desktop).

Once we extract the Fiji folder, we will see the new folder *Fiji.app*, which holds all necessary subfolders and files for Fiji. We don't have to run an installer; just download, extract the folder and start the application file *ImageJ - win64.exe* to run the program.

> **Attention!**
>
> You must have full write access for the location where you unzip the Fiji folder in order to have full functionality of ImageJ/Fiji.

Add the Plug-in for iPhone and Android phone videos

Before we can import videos from cell phones into ImageJ/Fiji, we must follow these steps in ImageJ/Fiji:

- Go to menu: Help → Update.
- Wait until it checks for updates.

- Click on the button on the bottom left "Manage Update Sites."
- Then a list opens up.
- In that list scroll down and put a check mark at the item "FFMPEG."
- Close that window by clicking on the button "Close."
- Confirm the selection in the next window by clicking on the button "Apply changes."

After these steps, we can open video files in the formats MOV (iPhone) and mp4 (Android) in ImageJ/Fiji.

4.2. Start ImageJ/Fiji

To start ImageJ/Fiji, we go to the Fiji folder and start the file *ImageJ - win64.exe*.

In lab courses, there might be a shortcut on the desktop (see figure below) which we double-click to start ImageJ/Fiji.

ImageJ-wi
- Shortcut

After starting the program, a small window from ImageJ/Fiji will open (see figure below).

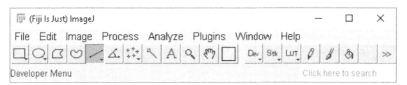

4.3. The Motion Videos

Recording our own Videos

We can take our own video of a two-dimensional motion using our cell phone's camera function or any other camera. We can then analyze the recorded motion using the video analysis program ImageJ/Fiji.

Examples of two-dimensional motions which we can analyze: a basketball throw, a pendulum, a person on a swing set, a running dog, etc. When we record the video, we have to make sure that we:

- support our camera to keep it still and that we don't follow the motion with our camera.
- film the motion so that we can observe the whole motion. That means the line of sight is perpendicular to the plane of motion.
- record the motion in a well-lit area.
- use objects which have a good contrast to the background and that are not too small. When we check our video, we should be able to see the objects.
- place a distance marker (e.g. an object which is 1 meter long) in the plane of motion, so it is visible in the video footage. We will use that to convert pixels to position data.

4.4. Import a Video into ImageJ/Fiji

To import a video from an iPhone or Android phone, we follow these steps:

- In the menu bar from ImageJ/Fiji we click on "File" → "Import" → "Movie (FFMPEG)..."

- Then a dialog window pops up:
  - Unselect "Use virtual stack."
  - To import the whole video, leave the default values for the first frame "0" and for the last frame "–1."

After clicking "ok" on the dialog window, the video opens up in a new window.

To play the video we click on the play arrow at the bottom left corner. We can also scroll through the video frame by frame using the slider in the progress bar underneath the video, or when we click on the arrows which are on the left and right ends the progress bar.

Rotating the Video

To rotate the video, go to the menu: "Image" → "Transform" → "Rotate 90 Degrees…"

4.5. Set up a Scale

To take actual position data from a video, we need to define the pixel-to-meter ratio. We do this by setting up a scale:

- On the menu bar on the top of ImageJ we click on the menu item for a diagonal line, which is the 5th symbol from the left (see circled item in the figure below).

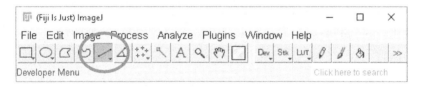

- We draw a line in a video frame over an object, whose length we know, by holding the mouse button while dragging the mouse.
- We click on "Analyze" on the top menu.
- Then we click "Set Scale" (see image below).

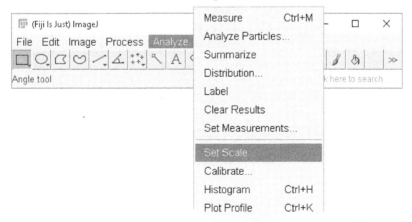

- In the dialog window which opens up we see the length of that line in pixels. We type in the real length of that line (see figure below). Then ImageJ/Fiji calculates and displays the ratio, how many pixels of the movie are how many meters in real life.

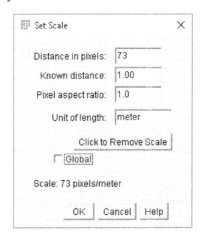

- Write down the scale information for your video as it is shown on your screen, for later reference. (Don't use the number from the image below, as this is just an example and varies from video to video.)

4.6. Data Collection

To collect motion data from the video, we follow these steps:

- From the top menu bar, we select "Plugins."
- Then we select "Manual tracking."

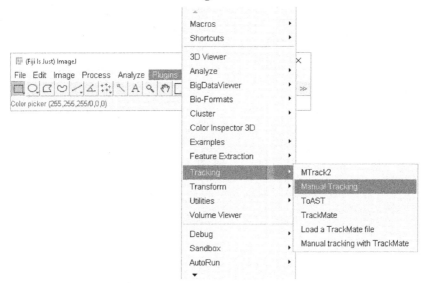

- After clicking, the "Tracking" window opens up (see figure below).

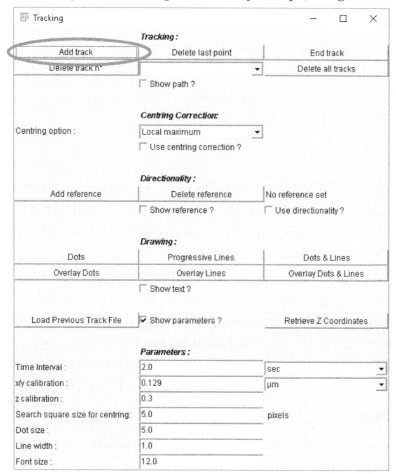

- To start the measurements, we click on "Add Track" (see circled button in the figure above). If we want to show the motion path, we check the box "Show path ?"
- Then in the video window, we click on the object of which we want to take the position data. By clicking on the object, we take its position data in pixels. After each click, the video moves automatically one frame forward. After taking the first position data, a new window with the table for the measurements opens up and will be filled out automatically with position data (see figure below).

Track n°	Slice n°	X	Y	Distance	Velocity	Pixel Value
1	57	390	462	-1.000	-1.000	13083806
1	58	392	450	1.569	0.785	12623255
1	59	392	438	1.548	0.774	12162449
1	60	396	420	2.379	1.189	7425608
1	61	394	396	3.107	1.553	12164244
1	62	396	378	2.336	1.168	12887712
1	63	396	360	2.322	1.161	13875379
1	64	398	340	2.593	1.296	14072763

- We follow the motion of that object with the courser frame by frame and click each time. Each click, the pixel position of the mouse pointer is recorded in the spreadsheet (see figure above) and the video moves one frame forward.

- We save the results from the data. An easy way to save it is using the copy-and-paste function. We do that by dragging the mouse cursor over the numbers to be selected while holding the left mouse button depressed. When everything we want to use is highlighted, we right-click and select "Copy" in the pop-up menu. Then we open Excel and right-click on a cell in a new document and select "Paste" in the pop-up menu.

Show Motion Path

In the "Tracking" window (see page 43) we can also check the box "Show path ?" to plot the trajectory.

When the motion we want to analyze doesn't start right at the beginning of the video, we can jump to the spot where the motion starts using the progress bar below the video before we start collecting the data points.

Create Motion Data with ImageJ

To obtain motion data in SI units from the video, we use the spreadsheet ImageJ has created for us while we were clicking in each video frame on the object (see figure above). ImageJ created a spreadsheet with the recorded position data, separated by the x-component and the y-component. But the position is measured in pixels and the time data is still missing. ImageJ just lists them in the order we clicked.

Most information in that table is not relevant for us. For the motion analysis, we are only interested in the position values for X and Y and the "Slice n^0." The values we use for our motion analysis are marked in the figure below.

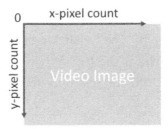

The X and Y values are the coordinates for the object in pixels. The "Slice n⁰" indicates from which video frame the position data was taken, which gives us the time information.

The orientation of the pixel coordinate system is shown in the image below.

Pixel coordinate system of ImageJ/Fiji

Analysis of Motion Data from the Video

Position Data

To complete the motion data from the spreadsheet ImageJ created for us, we still need to convert pixels to meters, and we have to add the time information.

We do both in Windows Excel. To import the data to Excel, we use the copy-and-paste function. In ImageJ, in the data table with the recorded measurements, we hold the left mouse button while we mark the part of the table we want to export. In ImageJ, we right-click and choose "Copy." Then we go to Excel and start a new file in which we paste the data into a new spreadsheet (press "Ctrl" and "V" at the same time). The result after importing the data to Excel is shown in the image below.

Now we clean up the table by removing columns and lines we don't need by highlighting the columns one by one by clicking on Excel's labels of the columns such as "A," "B," and so forth at the top. When a column we want to delete is highlighted (like column "E" in the example from the figure above), we select delete in the top menu.

We delete all columns except for the columns "Slice n⁰," "X," and "Y." The columns "X" and "Y" contain the position data (still in pixels, though). The column "Slice n⁰" contains the time information measured in number of video frames elapsed.

To convert the pixels to meters, we use the scale we determined for our video in a previous step (see section 4.5).

We create a new column for the position data in meters in Excel and define the table entries using a formula for these columns. For example, the column "x-position in m" is defined by

$$\text{x-position in m} = \frac{X}{Scale}$$

where X is the position in pixels and *Scale* is the ratio of pixels per meters.

To define a new column in Excel using the above formula, we highlight a new column by clicking on one of Excel's column labels. Then we type in the formula in the formula field as shown in the figure below. The formula has to start with "=" followed by the mathematical expression for the column. To input values from other columns in that formula we click on the respective columns. The entry should look like the one in the figure below, with different numbers, though.

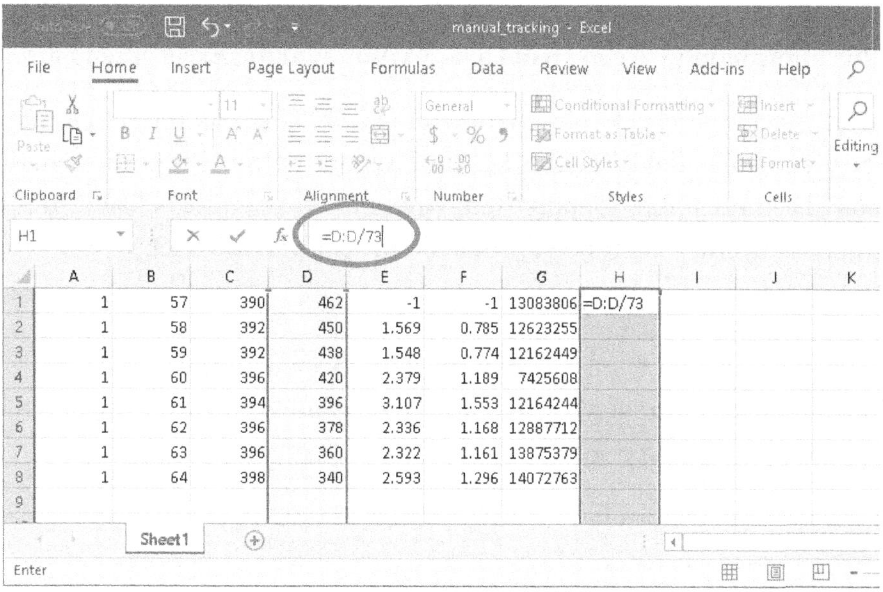

In the above example the position data in pixels is listed in column "D." To compute the position in meters in a new column (column "H" in this example) we type into the formula bar: "=D:D/73," where 73 is the scale for the video in this example. Scales are used to convert the pixel values into meters. In this example the scale is 73 pixels/meter. That means, the video has 73 pixels per 1.00 real-world meter.

To convert every pixel value into meters, we use the above formula for every field in the new column. To do that we click on the little square in the bottom right corner of the first cell of the new column and drag the mouse all the way to the bottom of the column while holding down the mouse button.

> **Attention!**
>
> We must use the scale from *our* videos. So, instead of 73, like in the above example, we plug in the scale in the formula of our own video. See section 4.5 on how to set up a scale.

Time Data

To get the time values, we use the frame number of the video. A typical video has 30 frames per second. That means each frame increases the time by 1/30 seconds. With this information we can create a new column in the Excel spreadsheet.

However, we should check the video's frame rate. To check the frame rate, right-click the video file in the Windows file explorer and select "Properties" in the pop-up menu. Then open the "Details" tab, where the frame rate is listed. It is easiest to just manually enter the time values in a new column in Excel.

Calculating the Speed through Video Analysis

To determine the speed of the motion shown in a video, we look at the x-component and y-component of the motion separately. We first convert the pixel information to meters (see section "Position Data" on page 45).

To get the x-component of the velocity we use

$$v_x = \frac{\Delta x}{\Delta t}$$

where Δx is the displacement in the x-direction and Δt is the time between the video frames where the displacement took place.

To get the y-component of the velocity we use

$$v_y = \frac{\Delta y}{\Delta t}$$

where Δy is the displacement in the y-direction and Δt is the time between the video frames where the displacement took place.

The two-dimensional velocity vector is

$$\vec{v} = \vec{v_x} + \vec{v_y}$$

The overall speed is

$$v = \sqrt{v_x^2 + v_y^2}$$

5. Creating Measurement Graphs with Excel

This chapter contains step-by-step instructions for creating graphs with Microsoft Excel 2013. Before we start with graphs, we learn how to create a new quantity in Excel using a formula.

5.1. Creating new Variables and Quantities Using Formulas

Sometimes we don't want to plot the measurement value, but a quantity which depends on that quantity. Before we can plot the new variable, we need to calculate it and list it in a new column in our data spreadsheet. In Excel we can enter formulas with inputs from other table columns to express dependencies on other columns or entries.

Below are some typical applications, where we define and calculate a new quantity or variable:

- Correcting the offset from position or time values:
  We create a new table column with values where we subtract the initial position or time value from each original value to start the motion at x = 0 and t = 0.
- Analyzing the kinetic energy of a system:
  When we are interested in the kinetic energy, we have usually measured the speed of the object. We then use the formula for the kinetic energy to create a new table column with the energy values, which depend on the speed values.
- Changing units:
  If we have measured the mass in grams, but for the analysis it would be useful to use kilograms instead, we can create a new table column with values in kilograms. We define the kilogram entries by dividing the original mass values by 1000.

Let's look at an example, where we start with two columns in an Excel spreadsheet.

To create a new table column for the calculated variable:

- We click on a new field of a new column, in which we want to add the new quantity.

- We then use the formula bar to type in the formula (see circled field in the figure below). Formulas start with an "=".
  To input values from other columns in the formula, we can click on the respective column label. This way, the calculated values correspond to inputs from the same row.

	A	B	C	D	E
1					
2			1	1.1 C)^2	
3			2	2.3	
4			3	6.5	
5			4	8.1	
6			5	9.5	
7					
8					

  Formula bar: SUM =0.5*(C:C)^2

- Next, we apply the formula to all fields in the new column. We do that by using the small square in the bottom right corner (see circle in the figure below) of the field we just entered and dragging the mouse to the end of the column.

	A	B	C	D	E
1					
2			1	1.1 C)^2	
3			2	2.3	
4			3	6.5	
5			4	8.1	
6			5	9.5	
7					
8					

  Formula bar: SUM =0.5*(C:C)^2

- Excel automatically applies the formula to compute the column's entries. The formula remains visible and editable in the formula bar.

5.2. Creating Graphs with Excel

We can use Microsoft Excel or similar software to create measurement graphs. In this section we will create the graph from the example from section 1.5. To create a graph, we enter the measurement values into a table in Excel as shown in the figure below on page 51.

When the table is complete, we follow these steps to create a measurement graph:

- We mark the complete columns of the table with the mouse by holding the left mouse button while dragging over the table.
- We click on "Insert" in the top menu to open the insert tab.
- We select "Scatter" from the "Charts" group (see circled item in the figure below).

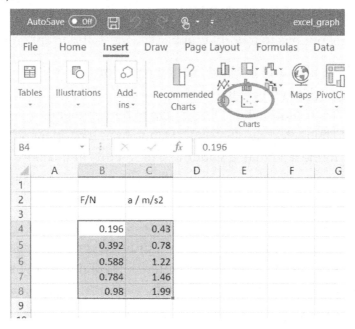

- Under the scatter graph menu, we can choose between several scatter graph types (see figure below). The first option works best to represent physics data.

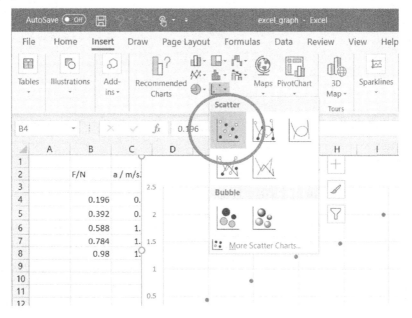

- After we have clicked on the graph type selection, Excel creates the graph with the first column on the x-axis, and the second column on the y-axis. If our table is horizontally oriented, Excel plots the first row on the x-axis and the second one on the y-axis of the graph. Excel automatically uses proper max values for both axes to show all data.

Adding Labels and Modifying the Graph

We have several options to modify graphs. The most important ones are:

- Switching the horizontal axis with the vertical axis
    - Click anywhere in the chart.
    - Then choose the "Design" tab in the top menu.
    - Then select the "Select Data" menu in the top menu.
    - In the new dialog window select "Edit" and choose the columns or lines for the axes manually.
- Adding labels for both axes
    - Click anywhere in the chart.
    - Click the "Chart Elements" button next to the upper- right corner of the chart. (See circled button in the figure below.)
    - Check the boxes in front of the elements you want to add from the menu.

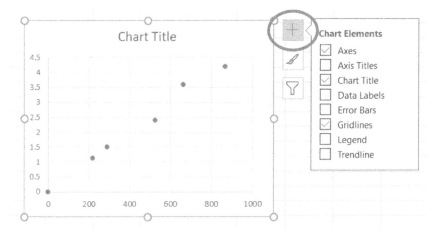

5.3. Adding Regression Lines and Curves in Excel

We can choose between fitting a line to our data points or fitting a curve to our data points. These lines or curves are called regression lines or regression curves. For the general physics lab course, the most relevant fit lines and curves are linear, quadratic, and exponential.

Linear Regression

To add a linear regression to our graph, we just check the "Trendline" option in the "Chart Elements" list (see image above).

Display the Equation for the Regression line

To display the equation for the regression line, we double-click on the line in the graph to open the "Format Trendline" pop-up menu. When we scroll down in that menu, we see the option to display the equation. (See image below.)

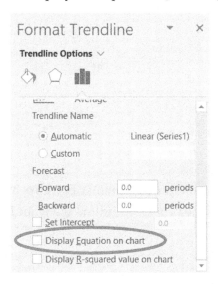

Quadratic and Exponential Regression Curves

To get an exponential or a quadratic Regression (which is called polynomial fit of the order of 2 in Excel), we do the following:

- We right-click on any data point in the graph and select "Add Trendline…" in the pop-up menu (see image).

- Then the "Trendline Options" menu opens up, and we select the radio button in front of the type of trendline we want Excel to calculate (see figure below).

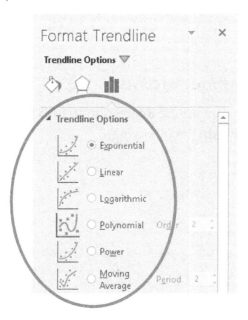

> **Attention!**
>
> If the variable for the vertical axis has a zero or a negative entry, Excel doesn't let us select "Exponential" because of the mathematical limitations of the exponential function. If we want to force Excel to calculate an exponential regression anyway, we change the "0" to "0.001" in the table, or delete these data

Display the Equation for the Regression Curve

To display the equation for the regression curve, we double-click on the line in the graph to open the "Format Trendline" pop-up menu. When we scroll down in that menu, we see the option to display the equation. (See image below.)

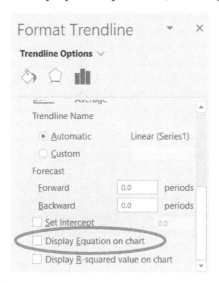

6. Creating Measurement Graphs with Logger Pro

When we take measurements with Logger Pro, the program automatically creates the measurement graphs of our data. The graphs show all our data, but this means that an outlier might cause the axis' scale to be way off, and the actual measurement data is hard to read. To correct the display of our data, we have to change the scales of the axes.

6.1. Zoom in or out in the Graphs

To zoom in our out, the easiest way is to just click on the current maximum value of the axis we want to change (see figure below). When we type in the new max value and hit enter, the graph automatically adjusts to the new max value.

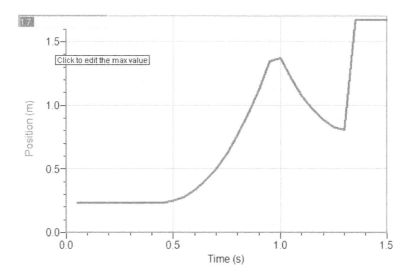

6.2. Adding Regression Lines and Curves in Logger Pro

Linear Regression

To fit a linear regression to our measurements, we mark the part of the graph for which we want to get a linear fit by holding the mouse button while dragging the mouse cursor. Then we click "Analyze" in the top menu. Then we choose "Linear Fit" (see the figure below).

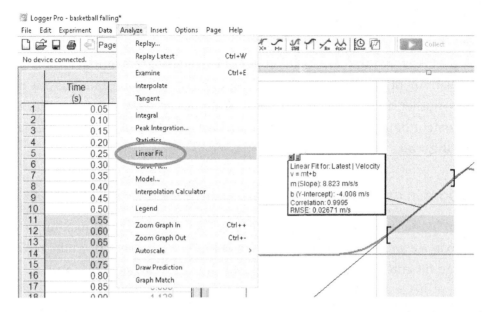

When Logger Pro calculates a linear fit for our measurements for the marked interval, a little box appears with the mathematical equation of the linear fit. See chapter 7 about the data analysis for further explanations.

Quadratic and Exponential Regression

To create a quadratic or exponential curve fit for our measurement data, we mark the part of the graph on which we want to get the fit by holding the mouse button while dragging the mouse cursor. Then we click "Analyze" in the top menu (see the figure above). Then we select "Curve Fit…" and then in the submenu we check the radio button to choose "Quadratic" or "Natural Exponent" (see figure below).

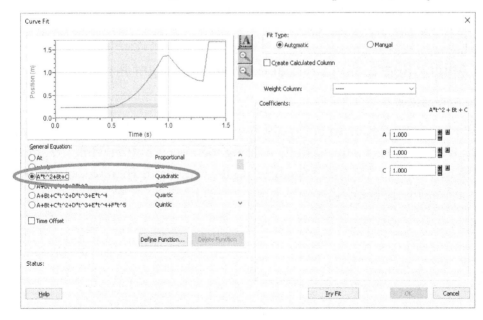

7. Data Analysis

When we analyze our measurement data and report our findings, we typically perform the following tasks:

- Creating measurement tables
- Cleaning up our data (deleting outliers)
- Comparing our results with the true values known from the literature
- Creating measurement graphs
- Analyzing the graphs and creating linear, quadratic, or exponential fit lines and curves
- Calculating averages of our measurements
- Summarizing the experiment
- Deducing or confirming a relationship between variables or a physics rule
- Discussing measurement errors

7.1. Clean up the Measurement Data

When we repeat measurements, we sometimes see outliers in our measurements, which are very different from the rest of the measurement values. Something must have gone wrong during that measurement and we should erase such data. When we use computer-based measurement systems, the computer often begins collecting data before the actual experiment has started. We can erase or ignore these parts of the data set as well.

7.2. Averages

When we repeat measurements, we see that the measured values fluctuate. The fluctuations are caused by measurement errors. When we take a measurement, the value is sometimes higher and sometimes lower than the actual value (see chapter 2). With repeated measurements, the errors level each other out somewhat in the long run. The average value of our measurement values will be a lot closer to the true value than each individual value alone. Thus, the average value is more accurate than each individual measurement.

There are several different types of averages. The most common is the arithmetic mean, which is often called just mean.

The mean is defined as

$$x_{mean} = \frac{1}{n} \sum x_{measured}$$

where $x_{measured}$ are the individual measurement values and n is the number of measurements.

Because of the statistical scattering of repeated measurements around the true value of the measured quantity, the error of the mean value is a lot smaller than the error of each individual measurement. The uncertainty interval for the arithmetic mean value is given by the standard error of the mean

$$SEM = \frac{SD}{\sqrt{n}}$$

where SD is the standard deviation or measurement error of the measurement values.

The true value of the quantity x_{true} is given by

$$x_{true} = x_{mean} \pm \frac{SD}{\sqrt{n}}$$

where x_{mean} is the mean value of the measurement values and n is the number of measurements.

That's why repeating the measurements and calculating the mean value produces more accurate results.

7.3. Comparing the Result with the True Value

When we have a measurement result, we like to compare our own result with the true value we know from the literature. To find out if our value makes sense, we also need to consider the uncertainty of our result (see section 2.1). As long the true value lies within the uncertainty interval of our result, we are satisfied.

7.4. Measurement Errors

Our experimental write-ups should discuss measurement errors and uncertainties (see chapter 2). We should mention the accuracy of the measurement instruments and methods, and possible origins of the errors. We can derive the limitation of the accuracy for our results from these errors.

A simple method to determine the actual measurement error is also to compare your result with the true value for that quantity we know from the literature. We can identify the standard error *SE* as

$$SE = |x_{true} - x_{measured}|$$

where x_{true} is the true value for that quantity from the literature, and $x_{measured}$ is the value we calculated or measured for that quantity.

Typically, we also want to calculate the difference between the two values in percent.

This can be done with the formula from section 2.1:

$$\%error = \frac{SE}{x_{measured}} \times 100\%$$

The formula tells us the percentage by which our measurement is off from the actual value.

> **Example**
>
> With our own measurements, we determined the gravitational acceleration to be 9.78 m/s² but the textbook says it should be 9.80 m/s². The percentage error is:
>
> $$\%error = \frac{\left|9.80 \frac{m}{s^2} - 9.72 \frac{m}{s^2}\right|}{9.72 \frac{m}{s^2}} \times 100\%$$
>
> $$= 0.82\%$$
>
> $$\approx 0.9\%$$
>
> Measurement errors are always rounded up to one significant figure.

7.5. Analyzing Graphs

Analyzing measurement graphs is a common method to investigate a physical fact or to find relationships between two quantities. Graphs also allow us to compensate for measurement errors by using fit lines or curves. The most common fits for measurements are linear, quadratic, and exponential. Chapters 5 and 6 give step-by-step instructions for creating regression lines and curves in both Logger Pro and Microsoft Excel. In this section we focus on how to choose the right fit for the measurement data, if linear, quadratic, or exponential, and what to do with these fit lines or curves. We will also discuss the accuracy of the fit lines.

Linear Regression

When we expect a linear dependency between two variables, or the measurement data points form an almost straight line, we can try to use a linear fit.

> **Example**
>
> When we walk at a constant speed, we expect the position x to change according to $x = vt$, where v is the speed, and t is the time, if the motion started at $x = 0$ at $t = 0$. We know that v is the slope of the position-versus-time graph.

For further analysis, we look at the mathematical equation for the fit line. In both Logger Pro and Excel, the linear fit equation has the form

$$y = mt + b$$

where y is the dependent variable (in this example the dependent variable y is the position x), m is the slope, t is the independent variable (in this example the time), and b is the offset.

The programs compute the coefficient m and the constant b so that the regression line best fits our data.

Quadratic Regression

When we expect a quadratic dependency between two variables, or the data points form a parabola-shaped curve, we can try to fit a quadratic regression.

> **Example**
>
> When we take position measurements for a free fall, we expect the position to change quadratically with the time. We know from the textbook that the distance traveled is $d = \frac{1}{2} a t^2$, with t being the time and a the acceleration, if the motion starts at rest at $d=0$ at $t=0$.

For further analysis, we look at the mathematical equation for the curve fit. In both Logger Pro and Excel, the quadratic curve fit equation has the form

$$y = At^2 + Bt + C$$

where y is the dependent variable, t is the independent variable, and A, B, and C are the computed coefficients.

> **Attention!**
>
> When we want to find out the acceleration of a motion with constant acceleration (like the free fall), we expect the position to change according to $x = \frac{1}{2} a t^2$, with a being the acceleration. This is, why we want to use the quadratic fit for the position-versus-time measurement data. However, the software calculates the coefficients for the standard quadratic formula $y = At^2 + Bt + C$. Thus, the calculated coefficient for A is $A = \frac{1}{2} a$. The acceleration a is $a = 2A$.

Exponential Regression

When we expect one variable to have an exponential dependency on another, or the data points form an arch starting flat and curving up or starting high and curving down, we can try to fit an exponential regression.

> **Example**
>
> When we take temperature measurements of a hot cup of water and let it cool, the temperature changes exponentially. We expect the temperature T to follow the function $T(t) = T_s + T_0 e^{-kt}$, where t is the time, T_0 is the initial temperature at $t=0$, T_s is the surrounding temperature, and k is a constant depending on how fast it cools.

For further analysis, we look at the mathematical equation for the curve fit. In Logger Pro, the exponential curve fit equation has the form

$$y = A e^{-Cx} + B$$

where y is the dependent variable and x the independent variable. The coefficients A and B and the exponent C are calculated through the fitting process.

In Excel, the exponential curve fit equation has the form

$$y = A e^{-Cx}$$

where y is the dependent variable, and x the independent variable. The coefficient A and the exponent C are calculated through the fitting process.

Excel doesn't calculate or allow the additional coefficient B as Logger Pro does (see above) for an offset of the values.

Accuracy of the Regression Lines and Curves

When we created the measurement graph, we repeated the measurement several times. In section 7.2 we saw that the accuracy improves when we repeat measurements more often and calculate the mean. When we use the coefficients of the regression line or curve to derive a result, it is also more accurate than each individual measurement alone.

How Good was the Chosen Model for the Fit?

When Excel or Logger Pro compute the regression lines or curves, the programs also tell us how well they match the data. The measure of how well the regression matches the data is the "correlation" in Logger Pro, or R^2 in Excel.

Logger Pro:

If the correlation is close to 1 or −1, then the model is a good fit. If the correlation is 1 or −1, then the model is a perfect fit. If the correlation is close to 0, then the chosen model is not a good fit and there is no correlation between the two variables.

Excel:

If R^2 is close to 1, then the model is a good fit; if R^2 is equal to 1, then it is a perfect fit. If R^2 is close to 0, then the chosen model is not a good fit and there is no correlation between the two variables.

Accuracy of the Coefficients of Fit Lines

For a more detailed look at the standard error of the computed coefficients we use a statistical function from Excel. We follow these steps to calculate the standard error for the computed value of the slope of the linear fit with Excel:

- We open the "Data" tab from the top menu.
- We select the "Data Analysis" function from the Excel menu (see figure below).

  o Adding the Data Analysis tool if the function is not in the menu: If the Excel top menu doesn't show that button, we have to add that function by opening the File tab, and under that tab we select "Options" → "Add-ins" and click "Go." Then in the pop-up menu we check the box for "Analysis ToolPak – VBA."

- After clicking the "Data Analysis" button, a pop-up menu opens, and we select "Regression" and hit "OK."

- In the next pop-up menu, we populate the fields for the inputs for x and y, by highlighting the respective columns or rows in the Excel spreadsheet.

After these steps Excel displays the statistical analysis of the measurement data, which should look like the following figure. The circled entry is the standard error of the slope, which is the uncertainty of the slope. This is the only statistical output we use for our analysis.

	Coefficients	Standard Error	t Stat	P-value	Lower 95%	Upper 95%	Lower 95.0%	Upper 95.0%
Intercept	0.005086913	0.295025133	0.0172423	0.9870691	-0.8140342	0.824208	-0.814034175	0.824208001
X Variable 1	1.996062239	0.114424625	17.4443415	6.34E-05	1.67836855	2.31375593	1.678368548	2.313755929

Part C

Lab Experiments

Lab 1: Measure and Describe Motions

Objective In this lab you will learn how to make your own motion measurements and how to create and read motion diagrams. You will also learn how to interpret motion diagrams, e.g. if the motion was forward or backward, or slow or fast.

Main Goal Learn to read and create motion diagrams for a walking person.

Pre Lab Assignment
1. Draw an example position-versus-time graph for a motion.
2. What is the formula for velocity?
3. Give the length in meters without using a prefix for the following lengths: 3 µm, 0.5 mm, 123 km.
4. Convert the following lengths into SI units: 1.0 inch, 0.25 inch, 0.125 inch.
5. What is the difference between instantaneous and average velocity?

Equipment LabQuest 2, motion detector, computer with Logger Pro software

Experimental Setup

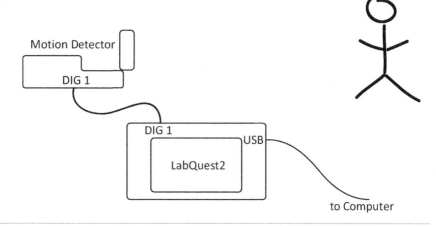

Tasks and Procedures
1. Take measurements of the motion of a walking person and plot the position-versus-time graph. Analyze normal walking, using both the position graph and the velocity graph (see page 31 for how to use the sensor):

- Create a position diagram.
- What is the average walking speed?
- How long does it take to reach the walking speed?

2. Plot a position diagram of the motion of a walking person with changing direction (just forward or backward, not sideways) and speed. Analyze this motion (using the position graph):
   - Calculate the maximum speed using the graph.
   - Create the velocity diagram.

3. Determine the maximal acceleration a human can have when she/he starts running. Show your data and your calculations.

Additional Questions

1. How can you determine the speed from a motion diagram?
2. How can you tell just from looking at a motion diagram if a motion was forward or backward?
3. When do you have negative values for the position?
4. What is the characteristic of a position graph for a motion with constant speed?
5. What does a position graph look like if the object is not moving?

Lab Report Grading

- Overall form (10%)
- Pre lab (10%)
- Task 1 (20%)
  - Motion diagrams (good data, labels, scale, units, zoom, etc.)
  - Calculations
- Task 2 (20%)
- Task 3 (10%)
- Additional questions (20%)
- Error discussion (10%)

Lab 2: Velocity and Acceleration

Objective In this lab you will learn about the behavior of accelerated motions. Specifically, you will be able to match position-versus-time graphs to velocity-versus-time graphs and to acceleration-versus-time graphs. Also, you will learn how to calculate velocities from position graphs, and how to calculate accelerations from velocity graphs using the graphs' slope. You will also learn to distinguish between uniform motions, constant-acceleration motions, and non-constant-acceleration motions.

Main Goal Investigate accelerated motions in terms of position, velocity, and acceleration and interpreting the respective graphs. Measure and estimate the influence of air drag.

Pre Lab Assignment
1. Look up the value for gravitational acceleration.
2. Calculate the speed of an object after 1.5 seconds in free fall, starting from rest (Neglect air drag.)
3. Calculate the percent difference, if a value changes from 12.5 m to 9.00 m.
4. What does terminal velocity mean in free fall?
5. How can you differentiate between uniform motions and accelerated motions? What information is necessary?

Equipment LabQuest 2, computer with Logger Pro software, motion detector, diverse objects to drop, misc. ring stand material (table clamps, metal rods, double clamps)

Experimental Setup

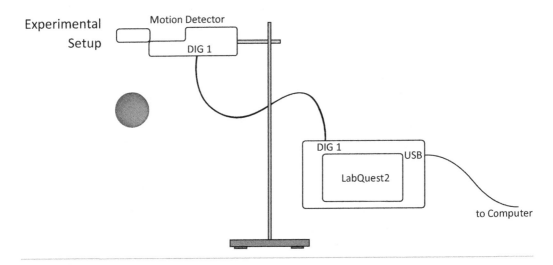

Tasks and Procedures

1. Take position, velocity, and acceleration measurements for motions of dropped objects (ball, crumpled paper, etc.), and a very loosely (almost flat) crumpled letter-size paper while they fall from a height of roughly 1.5 m. Analyze the motion of three objects using their graphs (see page 31 for how to use the sensor).

   - Plot a position graph for the three different objects (of which one is the almost flat, slightly crumpled sheet of paper).
   - Plot velocity and acceleration graphs for the motions.
   - Calculate the slope at one spot in each position graph and compare that result with the value you can read for that instance in the respective velocity graph.
   - Calculate the slope at one spot in each velocity graph and compare that result with the value you can read for that instance in the respective acceleration graph.
   - Discuss possible error sources. Which measurement data or graphs can you trust?

2. Determine the acceleration of a compact object (ball, crumpled paper) in free fall.

   - Using the position graph, you can create a curve fit in order to find the acceleration. Or, using the velocity graph, you can create a linear fit in order to find the slope.

- Calculate the percentage difference from the expected value.
3. Determine the terminal velocity of the very loosely (almost flat) crumpled paper.
4. Let a ball bounce off of the floor while you take measurements. Determine the ball's speed right before and after it bounces.

Additional Questions

1. Why is the acceleration you detected in this experiment positive? What does a positive or negative acceleration mean for a motion?
2. What type of motion do these objects perform? Use the graphs and what you know about uniform or constant-acceleration motions.
3. Could we approximately treat these motions as uniform or constant-acceleration motions? Maybe at least in parts?
4. For which object(s) is air drag negligible, and for which not?

Lab Report Grading

- Overall form (10%)
- Pre lab (10%)
- Position graphs (good data, labels, scale, units, zoom, etc.) (10%)
- Velocity graphs (good data, labels, scale, units, zoom, etc.) (10%)
- Calculated slopes (20%)
- Calculated gravitational acceleration and error percentage (10%)
- Determination of a bounced ball's velocity (10%)
- Answers to questions (10%)
- Error discussion (10%)

Lab 3: Two- and One-Dimensional Motion Using Video Analysis (2 Weeks)

Objective	In this lab you will learn how to analyze a motion using a video. You will learn how to determine the velocity one component at a time, and how to add the components together to a two-dimensional vector. You will also learn how motion analysis can be used to determine the force which accelerates the object.
Main Goal	Determine the acceleration of a motion using video analysis.

Pre Lab Assignment	1. Video analysis: If a video has 30 frames per second:
	   • How much time elapses after 1 frame?
	   • … after 5 frames?
	   • How many frames are in 1.5 minutes of video material?
	2. How do you convert the pixel information from the video into position data? Give an example. (See chapter 4 about video analysis.)

Equipment	Computer with ImageJ/Fiji software, cell phone or any other video camera

Tasks and Procedures	For the below tasks, please refer to chapter 4 on page 37 about video analysis.
	
	1. Record two types of motions using your cell phone or any other video camera:
	   • parabolic trajectory (e.g. basketball throw)
	   • a motion of an object or person, of which you determine the accelerations and forces which are involved (e.g. jumping up and landing on the ground, an object swinging on a cable, hammer landing on a nail head) through video analysis.
	
	When you take the video:
	   • Take the requirements of the video into account.

- Save the video file so that it is available during the lab course. (E.g. you can send the video file to your email-address or save it on a USB drive, which you bring to the lab.)
2. Plot the parabolic trajectory using the video analysis software.
3. Export the measurements from ImageJ/Fiji to Excel and edit the data. Create a table with position and time data (in SI units).
4. Analyze the velocity:

    Determine the two-dimensional velocity vectors at two different symmetrical instances.
    - Calculate the magnitude of the velocity vectors and the x and y components of the velocity.
    - Draw two-dimensional velocity vectors into the image of the trajectory. (You can use Paint, for example.)
    - Summarize your findings about the velocities.
5. Analyze the direction of the acceleration:

    Determine the direction of the acceleration in both videos and compare them. To determine the acceleration, you have to find $\Delta \vec{v}$. You need to find the change in velocity in a short time interval. So, don't just use the two velocities from before. For each instance draw two nearby velocities into one image. Then use the graphical method of subtracting the two vectors.
    - Draw the acceleration vectors into the image showing the trajectory.
    - Summarize your findings about the accelerations.
6. Investigate the occurring force during a motion in a real-world example.
    Help:
    Using the video, you can find out the acceleration. With F=ma you can determine the occurring force.

Additional Questions

1. What mathematical shape is the trajectory?
2. How are the x or y positions changing over time? Prove your statement.

Lab Report Grading

Week 1 (each week gets its own grade)
- Pre lab (10%)
- Task 1 (50%)
- Task 2 (20%)
- Task 3 (20%)

Week 2
- Overall form (10%)
- Task 4 (30%)
- Task 5 (20%)
- Task 6 (20%)
- Answers to questions (10%)
- Error discussion (10%)

Lab 4: Forces

Objective In this lab you will learn how to add forces. You will also learn how the gravitational force is related to the mass. And you will get to know what role forces play in real-world situations, when stopping a motion.

Main Goal Learn how to add forces (experimental and theoretical).

Pre Lab Assignment
1. List at least 7 different forces you know.
2. Split a vector with a length of 5 cm at a 45° angle to the x-axis into its x- and y- components.
   - Graphically
   - Mathematically
3. Explain the tip-to-toe method of adding vectors using an example.

Equipment Mechanical force gauges, LabQuest 2, force sensor, strings and rubber bands, mass pieces, scale, protractor, large paper, computer with Excel

Experimental Setup

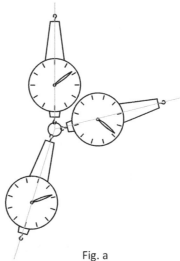

Fig. a

To determine the angles between the forces, it helps to draw the angle first on paper. Then hold the force meters above the drawing so that you match the angle.

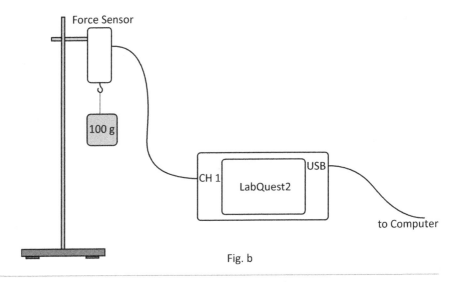

Fig. b

Tasks and Procedures

1. Using three manual force gauges, measure what forces are needed to have a force equilibrium when you add these three forces. Do two different set ups with different angles between the forces.

   a) One angle between two forces is 90°.

   b) Any different angles between the forces.

For both setups:

- Measure three forces, which pull all on one point. Combine forces of different values (see figure a of the experimental setup, and page 31 for how to use the sensor).
- Add the three force vectors graphically. Draw the force vectors to a scale, you choose.
- Add the forces mathematically using Pythagoras' theorem (if one angle is 90°).
- Compare the result with expected values and discuss possible error sources.

2. Investigate the relationship between the gravitational force and the mass.

   - For the investigation, take 5 measurement pairs of different masses and the occurring forces.
   - Plot the force versus mass in a graph, with force on the y-axis and mass on the x-axis (by hand or with Excel).
   - Determine the slope.

3. Investigate forces that stop an object's motion. Attach an object to a string, and the string to the force sensor. Lift the object a few centimeters and drop it. At the bottom, the object is stopped by that string. While the object is dropped and stopped, observe the occurring forces.

   - To begin, just hang the object on the string and measure the force.
   - Measure forces for different situations (use a rubber band instead of the sting, change the mass, drop from different heights, etc.). Compare the effect of the string type, while you keep the drop height and the mass constant.
   - Compare all measurements and try to find a pattern.

Additional Questions

1. What does the magnitude of the force in the experiment depend on, if an object falls and is stopped by a string or rubber band? Write down your observation.
2. Where does this effect play a role in the real world?

Lab Report Grading

- Overall form (10%)
- Pre lab (10%)
- Measured gravitational forces and masses (units, etc.) (10%)
- Measured forces at stopping an object on a string and rubber band (10%)
- Added forces measurements (20%)
- Added forces vectors graphically (10%)
- Added forces mathematically (10%)
- Answers to questions (10%)
- Error discussion (10%)

Lab 5: Newton's Second Law of Motion

Objective	In this lab you will learn about the relationship between force and the object's acceleration and mass.
Main Goal	Find Newton's second law. Investigate the relationship between the applied force and the acceleration and mass of the object.
Pre Lab Assignment	1. Explain in an example how you calculate the gravitational force of an object. 2. Sketch a graph where the variables x and y are proportional to each other. 3. Sketch a graph where the variables x and y are inversely proportional to each other. What are such graphs called?
Equipment	LabQuest 2, computer with Logger Pro software and Excel, motion detector, set of mass pieces, scale, track with cart and pulley wheel
Experimental Setup	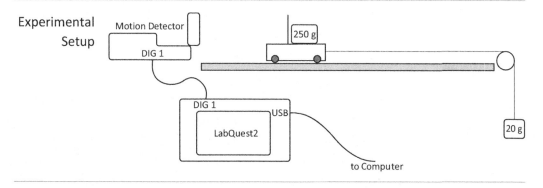
Tasks and Procedures	1. Find the relationship between force and acceleration (see page 31 for how to use the sensor): • Take motion data while the cart is accelerated along the track by the mass pieces pulling on it over the pulley wheel. Do five runs with different masses pulling on the cart. Vary the pulling mass by adding different amounts of 10 g pieces, starting at around 20 g. Keep the total mass of the cart constant during these 5 runs. Take two sets of measurements: one with the total mass of the cart around

0.75 kg and another one with the total mass of the cart above 1.00 kg.
- Calculate the gravitational force of the mass pieces for each run.
- Look at the velocity graph from Logger Pro for each run and determine the acceleration.
- Complete a measurement table of force and acceleration for each weight.
- Create a graph of the table from above (plot the force versus the acceleration).
- Analyze the graph and write down your findings.

2. Find the relationship between the mass of the cart and its acceleration:
   - Take motion data while the cart is accelerated along the track by the mass piece pulling on it over the pulley wheel. Do five runs for different cart masses (mass of cart alone is 500g + different amount of 250 g pieces added on the cart). The pulling mass on the pulley wheel must be constant (any mass between 20 and 100 g works).
   - Look at the velocity graph from Logger Pro for each run and determine the acceleration.
   - Complete a measurement table with the mass of the cart and its acceleration.
   - Create a graph of the table from above (plot a versus m).
   - Include a row in your measurement table which contains the product m×a (mass of cart multiplied by its acceleration).
   - Write down your findings and analysis of the graph.

Additional Questions

1. What did you find out about the relationship between force and acceleration? Write down your reasoning.
2. What did you find out about the relationship between mass and acceleration? Write down your reasoning.

Lab Report Grading
- Overall form (10%)
- Pre lab (10%)
- Measurements of force and acceleration pairs (10%)

- Force-versus-acceleration diagram (10%)
- Measurements of mass and acceleration pairs (10%)
- m-versus-a diagram (10%)
- Findings and reasoning (30%)
- Error discussion (10%)

Lab 6: Static and Kinetic Friction

Objective	In this lab you will learn about static and kinetic friction and find out what they depend on. You will measure friction and find the friction coefficient.
Main Goal	Measure the static and kinetic friction and calculate the friction coefficient.
Pre Lab Assignment	1. Draw a free-body diagram for the following situations: • Pulling a block sitting on a table with a small force, so that it doesn't move. • Pulling a block at constant speed across a table. 2. Explain, how you can calculate the friction coefficient if you know the normal force and the friction. (You can use an example.) 3. Why and how does water or oil affect the friction?
Equipment	LabQuest 2, computer with Logger Pro software and Excel, force sensor, set of mass pieces, scale, blocks to pull with different surface materials and sizes. For task 4, bring the material or thing you want to investigate (examples: friction of a shoe, rubber tire, rolling friction of a bike or skateboard).
Experimental Setup	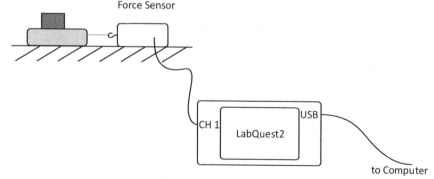
Tasks and Procedures	1. Pull on the force sensor that is connected to the block (wooden block with the wooden side down and an additional 0.5 kg

sitting on top of it). Observe the force which is applied to pull on the block (see page 31 for how to use the sensor).

- Increase the force gradually until it starts moving and then pull for another few seconds at constant force. Record the force data while doing that.
- Inspect the force versus time graph.
- Label the portion of the graph which is corresponding to the block at rest. Also, label where the block starts moving.
- Compare the two different surface materials of the block (using the surface area of the same size).
- Write down your findings.

2. Find the relationship between the mass of the object (block with different masses, sliding on the wooden and wide side) and the friction:

    - As shown in the experimental setup, you pull with the force sensor on the block by gradually increasing the force until it starts moving and then pull for another few seconds at constant force. Collect force data before and while the block is moving.
    - Run that procedure for several masses of the block (the empty block and pieces of different masses on top of the block). The mass pieces should have a mass of a couple of hundred grams each. Record the peak static friction and the kinetic friction for each run. Do three runs for each mass.
    - Create a measurement table which contains the total mass of the block, and the normal force on the block (calculated from the mass). For each setting record 3 different measurements for the peak static friction and for the kinetic friction. Report the average values for the friction as well.
    - Calculate the friction coefficient for each average peak static friction and average kinetic friction.

3. Observe and investigate the friction for different surface areas and materials:

- Do one run pulling the block with the thin side down and compare the measurement for static and kinetic friction with the measurements from part 1 with the wide side down. Also compare the friction forces which occur using the sides with the felt.

4. Perform an experiment about a real-world problem with the goal of finding the friction and friction coefficient. Also, discuss the implication of your experimental outcome on a real-world application.

Additional Questions

1. How does the size of the area affect the friction? Describe very briefly a possible experiment to investigate the influence of the size of the surface area on the friction.
2. Where does the difference between kinetic and static friction play a role in real life? Answer in a few sentences.

Lab Report Grading

- Overall form (10%)
- Pre lab (10%)
- Part 1 (20%)
- Part 2 (30%)
- Part 3 (10%)
- Additional questions (10%)
- Error discussion (10%)

Lab 7: Rotational Equilibrium and Torque

Objective In this lab you will learn about torque equilibrium. You will also learn about the law of the lever.

Main Goal Measure the distance to the pivot and the force and calculate the torque. Also solidify your understanding of torque equilibrium.

Pre Lab Assignment
1. What is the condition for rotational equilibrium?
2. Calculate the torque at the pivot, if a force of 53 N is applied perpendicular to a lever at a distance of 40 cm from the pivot.
3. A practical method to find the center of mass of an irregular-shaped object is to hang it at any point somewhere on the edge as shown on the figure below. Then you draw a line straight down from that point where it is hung up. The center of gravity is somewhere on that line. Explain why. What do you still have to do to find the center of gravity?

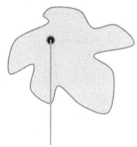

Equipment LabQuest 2, computer with Logger Pro software and Excel, force sensor, set of mass pieces, scale, meter stick, balance support, bar bell

Experimental Setup

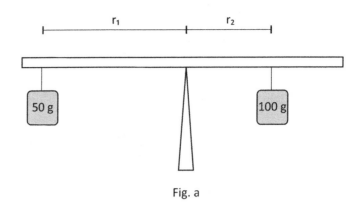

Fig. a

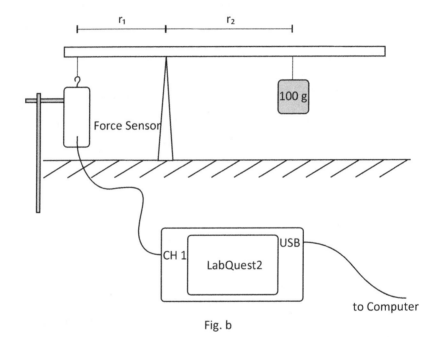

Fig. b

To balance the meter stick it helps if you place the support in the middle of the meter stick with the screw on the bottom.

Tasks and Procedures

1. Balance out a constant torque with different masses at different distances (see figure a for the setup):
   - Balance the empty meter stick exactly in the middle.
   - Hang a 100 g mass piece (the attached mass consists of the mass piece and the hook) on one side of the balanced meter stick roughly 25 cm from the pivot (this will be the side you leave unchanged).

- Balance out the torque from that 100 g piece by hanging different masses on the other side of the meter stick at different distances from the pivot.
- Record the following data:
  - Torque from the unchanged side with the 100 g piece
  - A measurement table for the other side with mass, force, distance, and force multiplied by distance (for at least 5 different settings)
- Analyze your findings (try to identify general rules or relationships).

2. Find the relationship between the distance and the forces at both sides of the horizontally balanced meter stick (see figure b for the setup; page 31 for how to use the sensor):

- Place the force sensor under one side of the meter stick at a fixed location of about 15 cm from the pivot. We call the distance between sensor and pivot r_{sensor}. The force sensor will be used to pull that side down while a 100 g piece is attached on the other side.
- Attach the 100 g piece (don't forget to consider the mass of the hook also) on the opposite side from the fulcrum and vary the distance to roughly 5 cm, 10 cm, 15 cm, etc. until 50 cm.
- Record the following data in a table:
  - Force at the sensor
  - Distance of the 100 g piece to the pivot (r_{100g})
  - $\dfrac{r_{100g}}{r_{sensor}}$
  - $\dfrac{F_{sensor}}{F_{100g}}$
- Relate this experiment to the law of the lever:
  $F_1 \cdot r_1 = F_2 \cdot r_2$ where
  F_1 and F_2 are two forces on opposite sides of a pivot (or fulcrum), r_1 is the distance between F_1 and the pivot (or fulcrum), and r_2 is the distance between F_2 and the pivot (or fulcrum).

3. Measure and calculate the gravitational torque of only the meter stick if the pivot point is 15 cm in either direction away from the center.
   - Use a mass piece to counter the torque from the meter stick (you choose the size and distance).
   - Calculate the torque from the mass piece.
   - Calculate the gravitational torque of the meter stick for that pivot. Do this by simplifying the object to a point-like object which is located at the center of mass.
   - Compare the torque you calculated for the mass piece with the one you calculated for the meter stick.
4. Measure, calculate, describe and explain the torques occurring in arm or leg exercises with a weight for at least 3 different situations.

Additional Questions

1. Where does the law of the lever play a role in real life? Find at least 3 examples.
2. Give an example related to your major where and how torque plays a role for an application.

Lab Report Grading

- Overall form (10%)
- Pre lab (10%)
- Task 1 (20%)
- Task 2 (20%)
- Task 3 (10%)
- Task 4 (10%)
- Additional questions (10%)
- Error discussion (10%)

Lab 8: Momentum and Collisions

Objective In this lab you measure speeds and masses of moving objects and calculate their momentums. You also determine the momentums of the individual objects as well as the total momentum of a system before and after collisions.

Main Goal Learn about momentums, elastic collisions, and inelastic collisions.

Pre Lab Assignment

1. An object (m = 500 g) with an initial speed of 0.2 m/s collides with another object (m = 1.5 kg) which was at rest before the collision. Calculate the resulting speed for an inelastic collision (when they stick together).

2. A small object (m = 200 g) collides elastically with a larger object (m = 1000 g), which was at rest before the collision. The incoming speed of the smaller object was 1.0 m/s. The speed of the larger object after the collision is 0.33 m/s.
Calculate the resulting speed and determine the direction for the smaller object after the collision when it rebounds. (Watch out for the directions of the motions and use respective signs for the velocities and momentums.)

Use the theory of conservation of momentum for your calculations.

Equipment LabQuest 2, computer with Logger Pro software and Excel, 2 photogates, track and carts with Velcro on one side and repelling magnets on the other side, extra masses, a light-blocking strip that can be attached to the carts

Experimental Setup

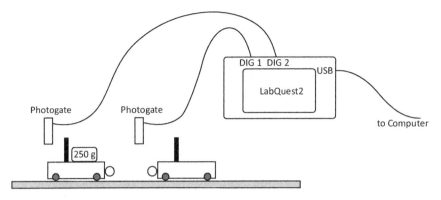

The photogates measure the speed of a passing cart. They can be used to measure the speed of several passing carts in a single experimental run.

Tasks and Procedures

See page 32 for how to use the sensors.

1. Find out through experimenting with elastic collisions (without taking measurements), in what situation the colliding cart
   - Rebounds back in the opposite direction
   - follows the other cart after the collision
   - comes to a full stop

2. Find out (or verify, if you already know it) the relationship between the total momentum of a system before and after two carts collide:
   a. elastically (when they rebound)
   b. inelastically (when they stick together)

   To counter the effect of friction, you can slope the track slightly.

   For both types of collisions (parts a and b from above) do the following:
   - Collide one cart with another on the track.
   - Give one cart a momentum through a push while the other one is at rest before the collision.
   - Take speed and mass measurements.
   - Calculate the total momentum before and after the collision.

- Create a table with speed and mass measurements and the calculated momentums.
- Vary both carts' masses.
- Analyze your findings (try to identify general rules or relationships).
- Suggested columns in the table with measurements and calculations: $v_{1,\text{before}}$, $v_{2,\text{before}}$, $v_{1,\text{after}}$, $v_{2,\text{after}}$, m_1, m_2, $p_{1,\text{before}}$, $p_{2,\text{before}}$, $p_{\text{total,before}}$, $p_{1,\text{after}}$ (if elastic collision), $p_{2,\text{after}}$ (if elastic collision), $p_{\text{total,after}}$

Additional Questions

1. What does this experiment have to do with traffic accidents?
2. Where else in real-life does momentum play a role?

Lab Report Grading

- Overall form (10%)
- Pre lab (10%)
- Task 1 (20%)
- Task 2 (40%)
- Additional questions (10%)
- Error discussion (10%)

Lab 9: Kinetic and Potential Energy Transformations

Objective	In this lab you will learn about energy transformation and energy conservation. You will learn how to design and perform an experiment to verify physics laws. You will learn to use the formulas to calculate an object's potential and kinetic energy.
Main Goal	Investigate the energy transformation from potential to kinetic energy or vice versa. Learn how to set up your own experiment.
Pre Lab Assignment	1. What are kinetic and potential energy and what are the formulas for these quantities? 2. Design an experiment where you show that when energy is transformed from kinetic to potential energy or vice versa that the energy is conserved. Available equipment: see below For manuals of the sensors see chapter 3; for video analysis, see chapter 4. • Make a list of the equipment you need for the experiment. • Make a detailed sketch of the setup. • Write down values of heights, masses, distances, etc. which are necessary to run your experiment. • Prepare an empty table with the quantities you want to measure. • Say what you want to calculate with the measured values.
Equipment	LabQuest 2, computer with Logger Pro software, video analysis program ImageJ/Fiji, and Excel, motion detector, force sensor, computer with Excel, track with cart, strings, mass sets, ring stand, pulley wheel, springs, scale, pendulum, photo sensor to measure speed, any other props you can bring to the lab
Experimental Setup	You design your own setup using the equipment from above.

Tasks and Procedures

Refer to chapter 3 for how to use the sensors; chapter 4 for video analysis.

1. Do three different versions of your experiment (different speeds or heights, etc.).
2. Compare the values for potential and kinetic energy before and after the energy transformation and write down your results.
3. Discuss the physics behind it and possible error sources and write it down.

Additional Questions

1. What could be the reason that in your experiment the values for potential and kinetic energy are not the same?
2. When you double the speed, what happens to the height?
3. What happens to the energy on a roller coaster?
4. Where else does the energy transfer from potential to kinetic or vice versa take place?

Lab Report Grading

- Overall form (10%)
- Pre lab/Experimental design and trial (10%)
- Detailed experiment description, so that someone who reads it can do your designed experiment without figuring out what to do and how to do it (you can refer to the experimental design) (10%)
- Measurement table organization, values, units, etc. (10%)
- Calculations for potential energy (10%)
- Calculations for kinetic energy (10%)
- Physics discussion of the experiment and your measurements (20%)
- Answers to the questions (10%)
- Error discussion (10%)

Lab 10: Specific Heat

Objective In this lab you will learn about specific heat, heat transfer, and how that is related to the temperature. You will see the effect of the specific heat of different materials on the temperature.

Main Goal Measure temperatures, calculate heat and the specific heat.

Pre Lab Assignment

1. How much thermal energy do you need to bring 0.5 liter of water to boil? The initial temperature is 70°F.
   a) Look up the formula for specific heat and temperature.
   b) Look up the specific heat for water.
   c) Convert the temperature into Kelvin.
   d) Calculate the heat.

2. Sketch the temperature curve of a cup of coffee while it sits in a room.

3. What is the difference between heat and thermal energy?

4. When you put a piece of metal into a cup of water, why is the magnitude of the change in thermal energy ΔE_{th} of the water the same as the change in thermal energy of the metal? But why is the change in temperature ΔT of the water not the same as that of the metal?

Equipment LabQuest 2, computer with Logger Pro software, temperature sensor, scale, calorimeter cup, heating device, ring stand, diverse pieces of different material (aluminum, copper, brass, and steel), water

Experimental Setup

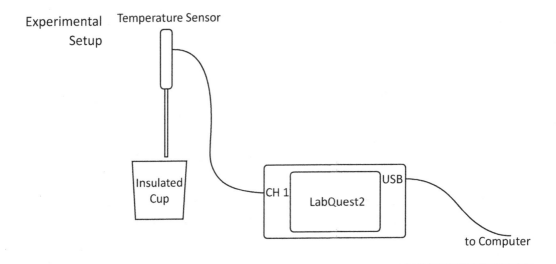

Tasks and Procedures

1. Take continuous temperature measurements of a hot cup of coffee (or water instead) while it cools. See page 35 for how to use the sensor.
   - Plot the temperature curve.
   - Find the mathematical function that fits the temperature trend. Identify the meaning of the coefficients in that temperature function.

2. Find the specific heat capacities of the different metals. Start with copper. For each metal:
   - You submerge small metal objects (one for each experiment) into hot water at a temperature of 170°F (≈350 K)
     - The metal should be at normal room temperature before you put it in.
     - Pour the hot water in the calorimeter
     - Don't use too much water when you fill the calorimeter cup! Use just enough water to fully submerge the metal piece in it.
   - Measure the temperature continuously with the LabQuest 2 and observe the temperature graph.
     - Start your continuous temperature measurements before you put the metal piece in. After roughly 30 seconds, put one of the metal pieces into the water while continuing the measurement. After you put the

metal piece into the water, keep measuring the temperature for another minute.
- Analyze the temperature diagram to find the temperature change caused by the added metal piece:
  To get the temperature change due to the added metal piece, you must interpolate the temperature backwards from when the metal piece has the same temperature as the water to the time exactly when you added the metal piece.
- Calculate Q, the heat which transfers from the water to the metal piece:
  How much heat Q did the hot water lose? You can calculate this by using $Q = mc\Delta T$, where $c = c_{water} = 4.184 \frac{J}{gK}$ is the specific heat of water. You get ΔT from the temperature graph. m is the mass of the water.
- Calculate the specific heat c of the metal:
  Using $c = \frac{Q}{m\Delta T}$, and the previous result for Q, you can calculate the specific heat of the metal.

Additional Questions

1. Why does a piece of metal at room temperature feel cold, but not piece of paper?
2. Where does the specific heat play a role in the field of your major?

Lab Report Grading

- Overall form (10%)
- Pre lab (10%)
- Task 1 (20%)
- Task 2:
  - Table with required data and measurements (10%)
  - Find temperature change (10%)
  - Calculations (20%)
- Answers to the questions (10%)
- Error discussion (10%)

Lab 11: Buoyancy

Objective In this lab you will learn about buoyancy and Archimedes' principle. You will also learn why boats float in water.

Main Goal Investigate buoyancy

Pre Lab Assignment
1. Draw a free-body diagram with all acting forces on an object when it is fully submerged in water. Draw another free-body diagram for a floating boat. Label all forces.
2. Look up Archimedes' principle and explain it using an example.
3. Bring to the lab any two different objects no larger than 4 × 4 × 4 cm (≈ 2 × 2 × 2 inches) that sink in water.

Equipment Force gauges or sensors with LabQuest 2, aluminum foil, aquarium, tape, strings, scissors, scale, overflow beaker, beaker, dish soap, sets of mass pieces to submerge in water

Experimental Setup

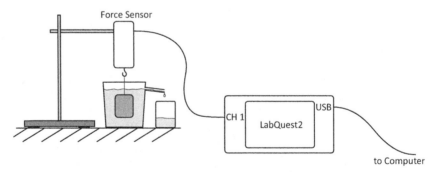

The overflow beaker has to be filled to the point of almost overflowing before the object is inserted. To remove the surface tension of the water, add a drop of dish soap.

Tasks and Procedures
1. Confirm Archimedes' principle ($F_b = \rho \cdot V \cdot g$) by measuring forces at fully submerged objects using:
   a) different materials of the same volume
   b) random objects, of different material, shape, and volume.

   See page 31 for how to use the sensor.

For both parts, a and b:
- Create a measurement table for your data keeping track of
  - volume
  - material
  - mass
  - gravitational force of the object in air
  - force down when submerged fully in water
- Submerge different objects in water and observe the displaced water and buoyancy force. The displaced water can be determined by one of the following methods:
  - using the overflow beaker
  - observing the rise of the water level in a container
  - if the objects has a simple geometrical form and sinks, using the mathematical formula for its volume
- Calculate the gravitational force of the displaced water
- Measure the buoyancy force indirectly by measuring the force the object is pulled down $F_{buoyancy} = F_g - F_{down}$
- Write down your findings

2. Investigate a floating boat (version a or b)

Make sure, the boat stays dry in the inside. Otherwise you are not able to measure how much water the boat displaces correctly.

a) Mini boat or container floating in a beaker

b) Boat floating in an aquarium

For both parts:
- Build a boat with the aluminum foil or use a container which floats on the water.
- Put some weights into the boat (it should still float)
- Calculate the gravitational force of the boat's mass including its load
- Measure the mass and volume of the displaced water when the object is partly submerged
- Calculate the gravitational force of the displaced water
- Write down your findings. What factors determine how much load a boat can carry?

Additional Questions

1. Why is it easier for kids to learn swimming in salt water than in fresh water? Give reasoning based on physics laws.

2. Calculate how much force you need to hold a basketball underwater. Show your calculation. Look up facts and constants you need for your calculation.

Lab Report Grading

- Overall form (10%)
- Task 1 (40%)
- Task 2 (30%)
- Answers to questions (10%)
- Error discussion (10%)

Formulas and Constants

Motion

Velocity: $v = \frac{\Delta x}{\Delta t}$

Acceleration: $a = \frac{\Delta v}{\Delta t}$

The position x of a motion as a function of the time t with constant acceleration a:

$$x(t) = x_0 + v_0 \cdot t + \frac{1}{2}at^2 ,$$

where x_0 the initial position at $t = 0$, and v_0 the initial velocity at $t = 0$ are.

The velocity v of a motion with constant acceleration a is given through:

$$v(t) = v_0 + at ,$$

Where t the time, and v_0 the initial velocity at $t = 0$ are.

The fallen distance d during a free fall starting at rest is given through:

$$d = \frac{1}{2}gt^2 ,$$

Where g the gravity acceleration, and t the time are.

Force

Newton's second law

$F = ma$,
Where F the force on the object, m its mass and a its acceleration are.

Friction

Static friction (maximum value): $F_{static\ friction,\ max} = \mu_s F_{normal}$,
where μ_s the static friction coefficient and F_{normal} the normal force between the

two surfaces are.

Kinetic friction: $F_{kinetic\ friction} = \mu_k F_{normal}$,
where μ_k the kinetic friction coefficient and F_{normal} the normal force between the two surfaces are.

Gravitational force on an object: $F_g = mg$,
where m the mass of the object and g the gravity acceleration are (on Earth in Texas: $g = 9.80\ \frac{m}{s^2}$).

Buoyancy: $F_B = \rho_f V_f g$,

Where ρ_f the density of the displaced fluid, V_f the displaced volume, and $g = 9.80\ \frac{m}{s^2}$ are.

Rotation and Torque

Torque

$\tau = rF_\perp$,
where r is the distance between the point where the force F acts and the pivot. $F_\perp$ is the component of the force F, which is perpendicular to r.
$\tau = r_\perp F$,
where $r_\perp$ is the component of the distance between the point where the force acts and the pivot, which is perpendicular to the force F.

Moment of inertia

For a solid cylinder: $I_{cyl} = \frac{1}{2}MR^2$
For a hoop or thin-walled cylinder: $I_{hoop} = MR^2$
Solid sphere: $I_{solidsphere} = \frac{2}{5}MR^2$
Hollow thin-walled sphere: $I_{sphere} = \frac{2}{3}MR^2$

Energy

Potential energy: $U_g = mgy$
Potential energy of a spring: $U_{spring} = \frac{1}{2}kx^2$

Kinetic energy (translational): $K = \frac{1}{2}mv^2$

Rotational kinetic energy: $K_{rot} = \frac{1}{2}I\omega^2$

Heat: $Q = mc\Delta T$

Temperature Conversion

$T_{(K)} = (T_{(°F)} + 459.67) \times 5/9$,

where $T_{(K)}$ is the temperature in Kelvin, and $T_{(°F)}$ the temperature in degrees Fahrenheit is.

Constants

Acceleration on Earth (in Texas): $g = 9.80 \, \frac{m}{s^2}$

Water density $\rho = 1000 \, kg/m^3$ or $\rho = 1.000 \, kg/l$

Specific heat

Water: $c_w = 4190 \, \frac{J}{kg \cdot K}$

Iron: $c_i = 450 \, \frac{J}{kg \cdot K}$

Brass: $c_b = 380 \, \frac{J}{kg \cdot K}$

Aluminum: $c_a = 900 \, \frac{J}{kg \cdot K}$

Copper: $c_c = 390 \, \frac{J}{kg \cdot K}$

Acknowledgments

Thanks to all lab instructors for the physics laboratory course PHYS 1115 at Texas State University in the years from 2016 to 2019. The lab instructors gave me valuable feedback about the lab course and where and what students need help with. This was crucial for me when I got the task to redesign and modernize the lab two years ago. Thanks also for using my experiments, instructions, and manuals for the lab equipment in the lab with the students, and for helping to improve them.

The instructors' feedback on how it went and what could be improved was crucial to tweak the lab equipment manuals, the lab experiments and instructions. All this feedback and input helped me a lot to write this lab manual.

Thanks also to Jace Toronto, who was a thorough editor and helped me to polish the text.

Other Related Work

Math for General Physics: A short introduction, 2017, written by Elmar Bergeler.

About the Book *Math for General Physics*:

This book helps students who take an algebra-based general physics college course with the math they need in this course.

The author explains the math in an easy-to-understand manner using clear and simple language. The book covers fractions, solving equations, exponential notation, vectors, functions, and sine, cosine, and tangent. Each chapter contains examples and exercise problems. The solutions for the problems are at the end of the book.

Available at

https://www.amazon.com/dp/B06WGMVJ23

Made in the USA
Coppell, TX
12 February 2022